W9-CMD-607

contents

introduction

Habitats and Adaptations

Naturalists are scientists who study the way plants and animals adapt to and live within their habitats (the places where they live). These scientists spend much of their time in the field or in the laboratory observing. They record their observations in logbooks. Many times, a naturalist will observe a particular behavior and record it, then recognize it as important later. Having a logbook to record observations helps naturalists to notice patterns over the years.

The experiments in this book are designed to have you observe and record observations in a logbook just as a professional naturalist does. A logbook can be any kind of notepad or binder— it does not need to be fancy. Remember, if you make an observation, it is important to record it as soon as possible, or you may forget.

This book is divided into two parts. The first part is about adaptations. Adaptations are ways in which organisms, or living things, cope with their environments. There are two main types of adaptations: structural and behavioral. Structural adaptations are physical features that help a plant or animal survive. The shape of a bird's beak, the coloration of a flower, or the thick coat of fur on a mammal are all examples of structural adaptations. Behavioral adaptations are ways in which an animal behaves that help it survive. Behavioral adaptations include hibernating in the winter, feeding at certain times, or acting a particular way around another living thing.

The second part of this book is about niches. All organisms live in an ecosystem, which is a system of living things and their environment. The role that an organism plays in the ecosystem—what it eats, where it lives, what it does—is called its niche. Even though organisms share the environment with other organisms, each one has its own unique niche.

Many of the experiments in this book provide detailed instructions for getting started. Beyond that, they depend on your observations. Each activity has its own instructions and safety information. After completing these activities, you will have a better idea of how naturalists make discoveries about the world around them.

Psyched for Science

Super Science Projects About Animals and Their Habitats

Allan B. Cobb

the rosen publishing group's
rosen central
new york

To my mom and dad for always encouraging my love of science and nature.

Published in 2000 by The Rosen Publishing Group, Inc.
29 East 21st Street, New York, NY 10010

Library of Congress Cataloging-in-Publication Data

Cobb, Allan B.
    Super science experiments about animals in their habitats / by Allan B. Cobb. — 1st ed.
                p. cm. — (Psyched for science)
    Includes bibliographical references.
    Summary: Introduces the concepts of behavioral and structural adaptations in animals through hands-on investigations and other projects.
    ISBN 0-8239-3175-7
    1. Habitat (Ecology)—Experiments—Juvenile literature. 2. Animals—Adaptation—Experiments—Juvenile literature. [1. Habitat (Ecology)—Experiments. 2. Animals—Habitations—Experiments. 3. Animals—Adaptation—Experiments. 4. Experiments.] I. Title. II. Series.
QH541.14.C62 1999
590'.78 21—dc21
                                        99-042687

Manufactured in the United States of America

1 Keeping Warm

Being cold is not much fun. In the winter, when you go outside, you put on a coat to stay warm. If it is really cold, you might put on gloves and a warm hat. People wear clothes to keep them warm. Animals do not wear clothes, of course, so they have to have other ways to stay warm. Some animals, such as dogs and cats, have fur. In the winter their fur thickens, and they are said to put on a "winter coat." In the spring they shed their winter coats. If you look closely at the winter fur, you will see that it is much finer than their normal fur. The fine fur fills in the spaces between the normal fur to make the coat much thicker. The thicker fur acts as insulation to hold heat in and keep cold out. Animals that live in cold climates develop much thicker winter coats than their relatives in warmer environments do. This is an adaptation that animals developed for living in cold climates.

Some animals, such as birds, do not have fur. Birds depend on their feathers to keep them warm. If you have ever looked at a feather, you might have noticed how delicate it is and wondered how it could keep a bird warm. Feathers overlap one another to keep wind and water off the bird's skin, and they trap a layer of air underneath. Heat from the bird's body warms the trapped layer of air, and the feathers act as an insulating layer to separate the cold outside air from the warm air against the skin. Just as cats and dogs have fine fur to help keep them warm, birds have fine feathers. These fine feathers are called down. Down is a very good insulating material. In fact, people use it in many different products to keep warm. Down is common in sleeping bags, coats, blankets, and vests. For down to insulate properly, it must be able to trap and hold warm air. This is the reason that down coats and down vests are thick and puffy.

In this experiment, you will explore how insulation keeps an animal warm. You will want to have an adult help you gather the insulating materials for this experiment. Do not use house insulation, because the fibers in it may irritate your skin.

Keeping Warm

What You Need

- A large coffee can
- 2 small cans
- Insulating materials (such as feathers, shredded foam rubber, pillow fill, synthetic fill, shredded cloth, packing peanuts)
- A thermometer

What You'll Do

#1 Select one type of insulating material and place it in the coffee can until the can is half full.

#2 Fill the small cans with hot tap water and place one in the center of the large coffee can.

#3 Add more insulating material to the coffee can until it is filled to the top of the small can.

#4 With the thermometer, take a temperature reading from each small can and record it in your logbook. Record the data in a table similar to the one in the diagram. (You can photocopy this table. Make a separate table for each type of insulation.)

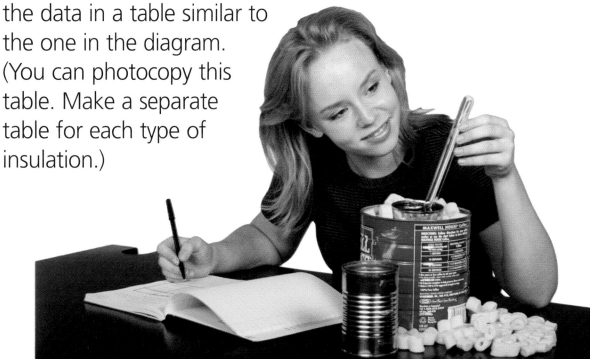

Keeping Warm

TIME	UNINSULATED CAN	INSULATED CAN MATERIAL:

#5 At ten-minute intervals, take the temperature of each small can. Be sure to let the thermometer stay in the can until it stabilizes, or gets a steady reading.

#6 Continue taking the temperatures for an hour.

#7 Repeat the experiment with as many different insulating materials as you have.

Analyzing Your Results

#1 Using the ten-minute intervals as the *x*-axis and the temperatures as the *y*-axis, graph the temperature over time. How can you tell from the graph which cooled fastest? Why do you think each material cooled at the rate it did?

#2 What do you think would happen if more insulation were packed into the coffee can? Experiment and find out.

#3 In the winter, when it is cold, do you think a thin blanket or a thick blanket is best for keeping warm? Do animals that live in cold places have thick or thin insulating layers of fur or feathers?

Keeping Warm

#1 Some animals, such as camels, have thick, woolly fur that helps keep them cool in the hot desert. Repeat the experiment using cold water instead of hot water. Were the warming curves similar to the cooling curves for each insulating material?

#2 Animals that live in extremely cold places, such as the Arctic, have thick layers of blubber to keep them warm. Blubber is a type of fat. Repeat this activity using solid vegetable oil—which is also a type of fat— as an insulating material.

#3 Some animals dig burrows into the ground to stay warm or to keep cool. Repeat this experiment using either soil or sand as an insulating material. Is the soil or sand better for insulation than the first materials you tested?

Dirt

Sand

2 Insect Camouflage

Some animals, including many insects, use a structural adaptation called camouflage. "Camouflage" means hiding by blending into the background. Some insects look just like the plants or landscape around them. This disguise is so good that birds often overlook these insects when searching for food. Some insects are shaped like leaves, and others have the same coloration as the plants on which they live.

In this activity, you will explore insect camouflage. Among insects there is incredible variety in size, shape, coloration, and habits. You will collect insects near your home and identify how (and whether) these insects camouflage themselves.

Be careful when collecting insects.

Watch out for stinging or biting insects. Do not handle insects with your bare hands, because they may bite. Using an insect guide, familiarize yourself with any dangerous or poisonous creatures that may live in your area, and know how to avoid them. You may want an adult to accompany you when you go out to a field.

Insect Camouflage

What You Need

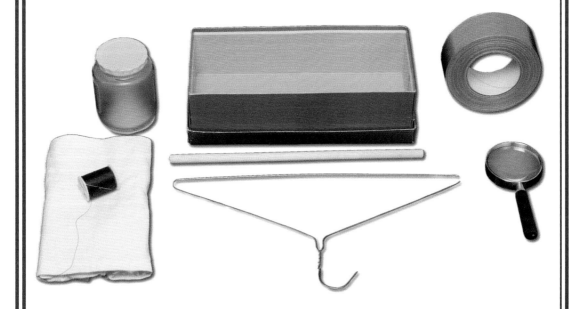

- A wire coat hanger
- A shoe box
- Jars
- Duct tape
- A broomstick or other long wooden stick
- A magnifying glass
- Cotton fabric
- A sewing machine or needle and thread
- An insect identification book (optional)

Experiment #2

What You'll Do

Getting Started

#1 Make a "crash net" for collecting insects by cutting the cotton fabric as shown in the diagrams.

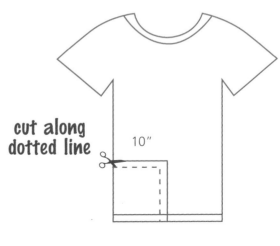

cut along dotted line

10"

#2 Have an adult help you sew the side seam of the net and a pocket around the edge.

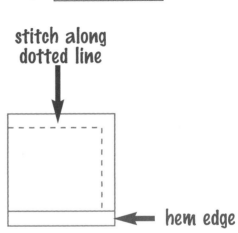

stitch along dotted line

hem edge

#3 Bend the coat hanger into a circle, slide one end into the pocket of the net, then attach the net to the handle as shown. Tape the wire to the handle.

Insect Camouflage

#4 Find an area with plants or bushes. You may want to wear protective gloves. Check the plants for insects. Look carefully—some may be camouflaged.

#5 Walk around and "crash" the plants with your net—gently, so you don't hurt the plants. Check to see if you have collected any insects. When you have caught several, put the jar over the insects, gently knock them into it, and quickly put on the lid. Be careful of stinging or biting insects. If you accidentally collect one, carefully remove it from your net and release it. Ask an adult to help you with this.

#6 After you have collected a number of insects, study them. Empty the net into the shoebox and sort the insects by types. Keep one of each type of insect and release the others.

Experiment #2

What You'll Do

Observation

#1 Observe the insects that you have collected. Make notes or sketches in your logbook for each insect. You may identify each one with your insect identification book, but this is not necessary.

#2 After completing your observations, release all of the insects.

Analyzing Your Results

#1 Do any of the insects have protective coloration or camouflage? What plants would they be camouflaged against?

18

Insect Camouflage

#2 How does this protective coloration help them?

#3 You may have encountered some insects that do not attempt to be camouflaged but are brightly colored and very visible. Why do you think that they do not attempt to hide?

For Further Investigation

Repeat this activity in another location that has different kinds of plants. Do you notice similarities or differences among the insects found and their camouflage patterns from one area to another? Why do you think they are the same or different?

3 Bird Beaks

Beaks are an important feature of birds. A beak may be long, short, fat, or hook shaped. A bird's beak is adapted to make getting food easier. The shape of the beak is a clue to what that bird eats.

Beaks come in a variety of shapes. Birds that eat seeds have short, stout beaks. Waterfowl, such as ducks, have flat bills that are used like strainers. Birds that catch fish by diving underwater have scissorlike beaks. Some birds that live near water have curved bills for scooping up small animals. Woodpeckers have long, sharp-pointed bills that they use to chisel insects out of tree bark and to carve nest holes in trees. The hummingbird has a long, thin beak that can reach deep into flowers to find the sweet nectar inside. These are all examples of ways in which birds' beaks have adapted over time to suit their various habitats and habits.

In this activity, you will have an opportunity to explore how different beak shapes are adapted to getting a particular type of food.

What You Need

- A bird identification book
- Food for the birds (Foods can include pop-corn, popped and unpopped; sunflower seeds; unshelled pecans or walnuts; shelled peanuts; uncooked rice; birdseed; raisins; and mini-marshmallows.)
- Chopsticks
- Needle-nose pliers
- Slip-joint pliers
- A slotted kitchen spoon
- A plastic straw
- A pair of tweezers
- A toothpick

What You'll Do

#1 Mix the foods that you collected in a bowl or dish.

#2 Try to collect food using each of the tools (the pliers, spoon, straw, tweezers, and toothpick), which represent beaks.

#3 Look through the bird identification book to find examples of birds that have beaks similar in shape, size, or design to the tools you used in your experiment.

#4 Make a data table similar to the one on the next page and record the foods that were most easily collected with each type of beak.

Bird Beaks

Data Table

Tool	Objects easiest to collect	Objects most difficult to collect	Example of a bird and what it eats

Analyzing Your Results

#**1** Which tools did you find the easiest for picking up small foods? Why?

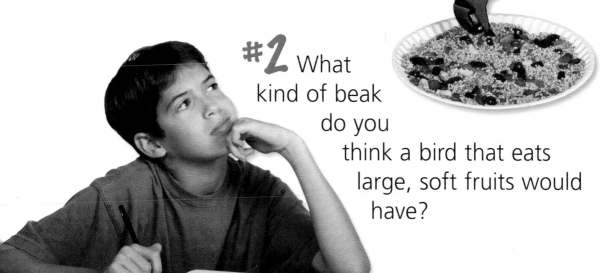

#**2** What kind of beak do you think a bird that eats large, soft fruits would have?

Experiment #3

For Further Investigation

#1 One of the earliest studies of bird beaks as adaptations was done by Charles Darwin on the finches of the Galapagos Islands. Research the different species of finches of the Galapagos Islands and identify how their beaks are adapted to the food they eat.

Finch bird

#2 Birds have beaks that they use to eat food, whereas other animals have teeth. Teeth can also give you information about an animal's diet. If you have a pet cat or dog, look at its teeth. Cats and dogs are carnivores, or meat eaters. How are their teeth adapted to a diet based on meat?

4 Feeding Time

Birds are everywhere. You see them along road-sides, in the country, and in the city. Birds are at home in a backyard or in the woods. Even in urban areas, a wide variety can be found. Birds are attracted to a number of different plants. Attracting wild birds is as simple as putting food out for them. You don't need a fancy bird feeder —even if you just spread food on the ground, birds will find it. (Feeders do make birds feel more comfortable, though.) On the ground, birds are easy prey for cats and other predators. The best place to set up a bird feeder is somewhere squirrels cannot reach. Squirrels will not harm birds, but they will eat their food.

In this experiment, you will build a simple bird feeder and stock it with different types of food. By carefully observing the birds that visit the feeder, you will soon determine which ones prefer which type of food. You will probably want an adult to help you construct the bird feeder and place it in a suitable location.

What You Need

- Four 16-ounce (.5-liter) plastic soda bottles or other small plastic bottles or jars
- A knife
- Waterproof glue
- A permanent marker
- A board measuring 18 x 1 x 6 inches (46 x 2.5 x 15 centimeters)
- Some heavy plastic
- A wire coat hanger
- Four different types of bird feeds (Use mixed birdseed, breadcrumbs, grains, etc.)
- A bird identification book
- Binoculars (helpful, but not required)

Feeding Time

What You'll Do

Getting Started

#1 With the knife, carefully cut the 16-ounce plastic bottle into four cups, each approximately 2 inches (5 to 6 cm) deep.

#2 Using the glue, attach the four cups to the board. Punch small holes in the bottom of each cup so that rainwater will not collect in them. Label each cup.

#3 Make a roof for your feeder from heavy plastic and a wire coat hanger.

#4 Find an isolated spot for the feeder. Choose a location where you will easily be able to observe it. The feeder should be near a tree or

shrub so that birds will have shelter to fly to if they are disturbed. The feeder should also be up off the ground. Don't set it on a tree limb, because squirrels will eat the food in it.

#5 Fill each of the cups with a different type of feed and write down which type is in each cup. Leave the feeder undisturbed, except to refill food, for a few days before beginning your observations.

Observations

#1 Observe your feeder as much as possible for two weeks. In your logbook, make notes about which birds eat from which cup. When you see a bird that you don't recognize, make notes about its appearance so that you can find it in a bird identification book.

Feeding Time

#2 Observe the feeder at as many different times of day as possible. In your log, record the time of day and note how much time you spent observing the feeder.

Analyzing Your Results

#1 What type of feed attracted the most birds? Why?

#2 Look back over your notes from this experiment and the one on bird beaks. Also look at the pictures of the birds in the identification book. How is the shape of the beak related to the type of food each bird chose?

#3 Continue the observations at your bird feeder throughout the year. It is important to keep refilling the feeder even if you are not making observations.

#4 Research the common birds in your area. Which birds migrate, and which ones stay in the area all year round? Do you live in a place that birds migrate through or migrate to? Where do the birds go when they migrate?

For Further Investigation

#1 Replace the feed in the feed cups with different foods. Try foods such as oats, cereals, suet, or dried fruit.

#2 Relocate your feeder or construct more feeders. Do you find that the birds prefer some locations more than others? If so, why?

5 Aquatic Habitat

In this activity, you will set up an aquarium and observe the habits of the fish that live there. An aquarium is a kind of mini-ecosystem. Each species of fish fills a niche within the ecosystem.

As you observe each fish, look for behavior that indicates the role it plays in its environment. For example, mosquito fish eat insect larvae and will stay close to the surface looking for them. Catfish are scavengers, which means that they eat what other animals discard. You will see catfish at the bottom of the aquarium, poking around in the gravel for bits of decaying matter.

If you have never set up an aquarium before, you will want to read through some books on setting up and maintaining an aquarium, such as those listed in For Further Reading at the back of this book. The best aquarium is one that is "balanced." A balanced aquarium is one that has enough plants to produce oxygen for all the fish. In this activity, you will use an air pump to supply extra oxygen.

Experiment #5

What You Need

- 10-gallon (38-liter) or larger aquarium with a lit hood
- Air pump, air stone, and filter (Some filters may take the place of the air pump and air stone.)
- Sand, small gravel, and several large, clean rocks
- Aquarium heater
- Thermometer
- Small fishnet
- Aquatic plants
- Several different kinds of fish
- Fish food
- A logbook

Aquatic Habitat

What You'll Do

#1 Set up the aquarium in a safe place where it is out of direct sunlight and away from a heater or air-conditioning vent.

#2 Place the small gravel in a bucket and rinse it with water to remove any dirt or debris. Pour about 1 inch (2 to 3 cm) of gravel into the aquarium. Smooth the gravel and create a slight upward slope from front to back.

#3 Place the sand in a bucket and rinse with water. Carefully pour the sand into the aquarium and smooth it out over the gravel.

#4 Place the rocks near the back of the aquarium.

#5 Fill the aquarium with water and set up the air pump, air stone, and filter. Run the air pump for 24 hours to get rid of any chemicals in the water before adding fish.

#6 Add plants to your aquarium.

#7 If needed, install the heater and adjust it to maintain the right water temperature for the fish that you have chosen. Once the temperature has stabilized, you are ready to add fish.

#8 Add the fish of your choice. If you choose to use fish from a nearby stream or pond, check local laws before collecting them. It may be illegal to take certain fish. If you buy fish at a pet store, check to make sure that all of them are compatible to live in the same tank. To help keep the aquarium clean, add several snails. Make sure to buy snails from a pet store rather than collecting them from outside, because some snails carry parasites that can infect fish.

#9 Feed your fish twice a day, morning and evening. Feed the fish only enough food so that they eat all the food in a five-minute period.

Aquatic Habitat

Analyzing Your Results

#1 In your logbook, record your observations about the feeding habits of different fish, the interactions among the fish, where they spend their time, where they hide, and so on.

#2 Fish have specialized mouths for feeding at the top, bottom, or middle of the body of water that they live in. In their natural habitat, fish that prefer to feed at the water's surface are called top feeders. Top feeders have mouths that open upward (rather than forward, the way yours does). Bottom feeders have mouths that open downward. In an aquarium, most fish feed at the surface because they have a regular feeding schedule; they expect to find food at the surface at regular times. Fish that live in large bodies of water eat whenever they encounter food or when they are hungry. Look at the fish in your aquarium. Based on the shapes of their mouths, what types of feeders do you have?

#3 The body shape of a fish can be a clue to its way of life. Fast-swimming fish have bodies that are streamlined all over, and fish that are capable of quick bursts of speed have bodies that are narrow from side to side. Fish that are adapted to living on the bottom of the water have bodies that are flattened. These are the three most common body types for aquarium fish. Look at the fish in your aquarium. Which group do they fall into?

For Further Investigation

#1 Choose one of your fish and observe how it uses its fins. Compare several different fish. Do they all use their fins the same way?

#2 You may want to put together more aquariums with fish from local streams and ponds. Set up one aquarium with organisms from a local pond and the other with fish from a local stream. Look for differences between the fish that come from ponds and those that come from streams. Why do you think these differences exist?

6 Predator and Prey

An animal that eats plants is called an herbivore, and an animal that eats other animals is called a carnivore. Animals that eat other animals are also called predators, and animals that are eaten are called prey. A single creature can be a predator of one animal and the prey of another. These relationships also exist among insects.

Both predators and prey fill important niches within an ecosystem. Prey insects are often considered pests because they eat beneficial plants, but they are an important part of the food chain. Predator insects are important because they control the population of pest insects. Sometimes when people spray plants with pesticides, the pesticides disrupt the balance between predators and prey. Because most pest insects reproduce quickly, if too many of their predators are killed off, their numbers explode, and they can cause a great amount of damage in a short time. Sometimes pest insects are controlled by purposefully releasing predator insects in the same area. In fact, many garden suppliers or nurseries sell praying mantises and ladybugs, both of which are beneficial predators.

In this experiment, you will visit a wilderness area or garden and closely examine the plants. When you find insects, you will collect them, identify them, and determine whether they are predators or prey. Be very careful when collecting insects. Watch out for ones that sting or bite. It is a good idea to wear gloves when handling insects, because they may bite. Familiarize yourself with any dangerous or poisonous animals that may live in your area, and know how to avoid them. You may want an adult to accompany you when you go out to a field.

Predator and Prey

What You Need

- Tweezers
- A magnifying glass
- A logbook
- A clear plastic or glass jar with a lid (for collecting insects)
- An insect identification book

Experiment #6

What You'll Do

#1 Find a suitable garden or nature area. Make sure to get permission from the owner.

#2 Choose a plant and search it for insects. Look carefully under the leaves, along the stem, in and around flowers, and on the ground at the base of the plant. Collect any insects that you find and place them in a jar. Be careful not to collect stinging or biting insects such as bees, wasps, or ants.

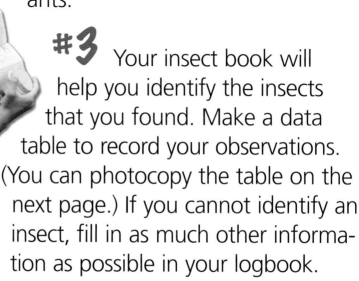

#3 Your insect book will help you identify the insects that you found. Make a data table to record your observations. (You can photocopy the table on the next page.) If you cannot identify an insect, fill in as much other information as possible in your logbook.

Predator and Prey

#4 Repeat the procedure with another plant. If the plant does not have any insects, note this in your logbook and select another one.

Analyzing Your Results

Insect Name	# of each collected	Location on the plant	Coloration	Predator or prey

#1 Did you find more predators or more prey? Why do you think this is so?

#2 Where did you find most of the predator insects? Where did you find most of the prey insects?

#3 How does coloration help predators and prey?

#4 Calculate the ratio of predators to prey for each plant you looked at. To calculate this ratio, write a fraction with the number of predators as the nominator and the number of prey as the denominator. For example, if you found 3 predator insects and 18 prey insects, write the fraction as 3/18. This fraction simplifies to 1/6. This tells you that for each predator, 6 prey insects were found.

#5 Compare the ratios of predators to prey for each plant of the same type that you inspected. Are the ratios similar?

#6 Total the numbers for all of the predators and prey for the area that you inspected and then calculate the predator-to-prey ratio. Is this ratio similar for each of the plants inspected, or did some plants have different ratios?

Predator and Prey

For Further Investigation

#1 Repeat this activity in another area with different types of plants. Compare the results of the two areas. Which area had a greater predator-to-prey ratio?

#2 Repeat this activity at a different time of year. For example, if you performed this experiment in the spring, repeat the activity in the summer, fall, and if possible, winter. How did your results change throughout the year?

glossary

adaptation A specialized method used by an organism for survival.

aquarium An aquatic habitat for fish.

aquatic Having to do with being underwater.

beak A structure forming the mouth parts of a bird; also called a bill.

behavioral adaptation The way an animal acts that helps it survive.

bill A structure forming the mouth parts of a bird; also called a beak.

camouflage Appearing to be part of the natural surroundings through similar coloration or shape.

carnivore An animal that eats other animals.

ecosystem An ecological community and its physical characteristics.

herbivore An animal that eats only plants.

insulation Material that prevents the passage of heat from one region to another.

naturalist A scientist who studies the relationships and interactions among plants and animals.

niche The role that an organism plays in an ecosystem: what it eats, where it lives, and what it does.

predator An animal that survives by eating other animals.

prey An animal that is eaten by other animals.

structural adaptation Physical characteristics that help an organism survive.

resources

Want to take the experiments in this book a step further? Maybe you would like to make a science fair project based on one of the activities in this book. Or maybe you just want to keep doing experiments about animals in their habitats. You'll find information, ideas, and inspiration at these Web sites:

Cool Science for Curious Kids
http://www.hhmi.org/coolscience

Cyberspace Middle School—Science Fair Projects
http://www.scri.fsu.edu/~dennisl/special/sf_projects.html

Exploratorium
http://www.exploratorium.edu

The Franklin Institute
http://sln.fi.edu

Mad Scientist Network
http://www.madsci.org

National Science Foundation Science in the Home
http://www.ehr.nsf.gov/ehr/ehr/science_home/html

Newton Ask a Scientist
http://newton.dep.anl.gov/aasquest.htm

The Science Club
http://www.halcyon.com/sciclub

Science Fair Project Ideas
http://othello.mech.nwu.edu/~peshkin/scifair/index.html

Scientific American Explore!
http://www.sciam.com/explorations

for further reading

Bailey, Jill, and Tony Seddon. *Mimicry and Camouflage*. New York: Facts on File, 1988.

Forsyth, Adrian. *Exploring the World of Insects: The Equinox Guide to Insect Behavior.* Buffalo, NY: Firefly Books, 1992.

Glass, Spencer. *Setting Up an Aquarium.* Neptune, NJ: T F H Publications, Inc., 1998.

Peacock, Graham, and Terry Hudson. *Habitats.* Austin, TX: Raintree Steck-Vaughn, 1992.

Peterson, Roger T. *A Field Guide to Insects*. New York: Houghton Mifflin Co., 1998.

Roth, Charles E. *The Amateur Naturalist: Explorations and Investigations.* Danbury, CT: Franklin Watts, 1994.

Savage, Stephen. *Animals Undercover.* New York: Simon & Schuster, 1994.

Weiner, Jonathan. *The Beak of the Finch: A Story of Evolution in Our Time.* New York: Alfred A. Knopf, 1994.

index

About the Author
Allan Cobb is a freelance science writer living in Central Texas. He has written books, radio scripts, articles, and educational materials concerning different aspects of science. When not writing about science, he enjoys traveling, camping, hiking, and exploring caves.

Photo Credits
Cover photo by Scott Bauer. P. 20 © Layne Kennedy/CORBIS; p. 24 © Kevin Schafer/CORBIS; p. 37 © Gail Shumway/FPG; P. 38 © Ralph A. Clevenger/CORBIS; All other photographs by Scott Bauer.

Design and Layout
Laura Murawski

Consulting Editor
Amy Haugesag

Metric Conversions
To convert measurements in U.S. units into metric units, use the following formulas:

1 inch = 2.54 centimeters (cm)	1 ounce = 28.35 grams (g)
1 foot = 0.30 meters (m)	1 gallon = 3.79 liters (l)
1 mile = 1.609 kilometers (km)	1 pound = 453.59 grams (g)